THE MATHEMATICS OF VIRUS SPREAD FOR BEGINNERS

An Introductory Work For The General Public

Jerald D. Truesdell

Copyright © 2020 by Jerald D. Truesdell
All rights reserved. This book or any portion thereof
may not be reproduced or used in any manner whatsoever
without the express written permission of the publisher
except for the use of brief quotations in a book review.

Printed in the United States of America

To my father, Donald W. Truesdell
who was an honest man

Acknowledgments

Special thanks to my precalculus class at Niagara County Community College, in the spring of 2020: Yara, Dawson, Stephen, Alexandra, Sandrae, Tanner, Rahshamel, Kaelie, Sanaa, Elise, Daniel, Sharon, Nabeel, Damien, Henessey, Justin, Kayleigh and Justin.

Thanks to my wife, Robin, who understands my eccentricities.

Thanks also to the late Dr. Paul "Doc" Kwitowski, who was a pedagogical genius.

v

Table of Contents

Chapter 1: Simple Exponential and Logistic Models………..……... 1

Chapter 2: Making Predictions………………….…............8

Chapter 3: What Mechanism Produces the Logistic S-Curve?...............15

Chapter 4: The Apex and S-Curve Basics……………………........... 26

Chapter 5: Estimating the Logistic Growth Rate When $I_0 \ll K$…........... 31

Chapter 6: Doubling Time and the Growth Rate………………….. 38

Chapter 7: What About R_0 and R_e?..43

Chapter 8: r_e and the Daily Growth Rate…………………….. 49

Chapter 9: Using R_e to Find r_e…………………………….............. 54

Chapter 10: The Effect of Social Distancing……………….............58

Chapter 11: The SIR Model Simply Explained……………………. 66

Chapter 12: Some Projections Based Upon the SIR Model……………78

Variables, Parameters and Formulas……………………….…....83

Answers…………………………………………………...85

Introduction

In spring 2020 a new language was cast upon us. Such terms as flattening the curve, exponential phase, apex and R-naught have become standard fare for the daily news. In this book we explain these quantitative terms.

In mid-March the whole State University of New York system was very quickly closed. Governor Cuomo acted very swiftly, and rightly so. Within a matter of hours we had to make a decision of how best to finish our classes at Niagara County Community College in Sanborn, NY. I am a professor of math and computer science, and this book is how I finished precalculus.

This brief text starts with exponential models, works its way through logistic models and then finishes with the SIR model for virus spread. Although some of the content is appropriate for 11th grade math and precalculus, there is still much to be comprehended without those corequisites. Most of this text can be read and comprehended without knowledge of logarithms.

We first consider an exponential model, and then show why it falls short. Then we move on to an improvement, which is the logistic model. That too is inadequate - although it would work well for a zombie apocalypse - and we explain why it falls short. Then we finish with the SIR model, which is a great fit for modeling virus spread.

The SIR model usually resides in the domain of differential equations and graduate-level epidemiology, which makes it inaccessible for the general public. We break it down for the general public in a very simplistic, straight forward way. All we need is arithmetic and a little patience to understand both the logistic and SIR models. Understanding the SIR model provides compelling reasons for following various public health initiatives during a pandemic. Understanding the SIR model will also dispel a lot of ignorance in the public square.

So, why all the numbers? The reason is simple: it is one thing to look at a graph, but quite another to look at a graph and understand the method that was used to produce that graph. With knowledge of the method comes a deeper level of understanding, as well as belief, or trust. A lot is to be learned by looking under the hood and gaining a true understanding of the engine that drive virus spread. Doing some calculations promotes understanding of the SIR model, or of any other.

Chapter 1: Simple Exponential and Logistic Models

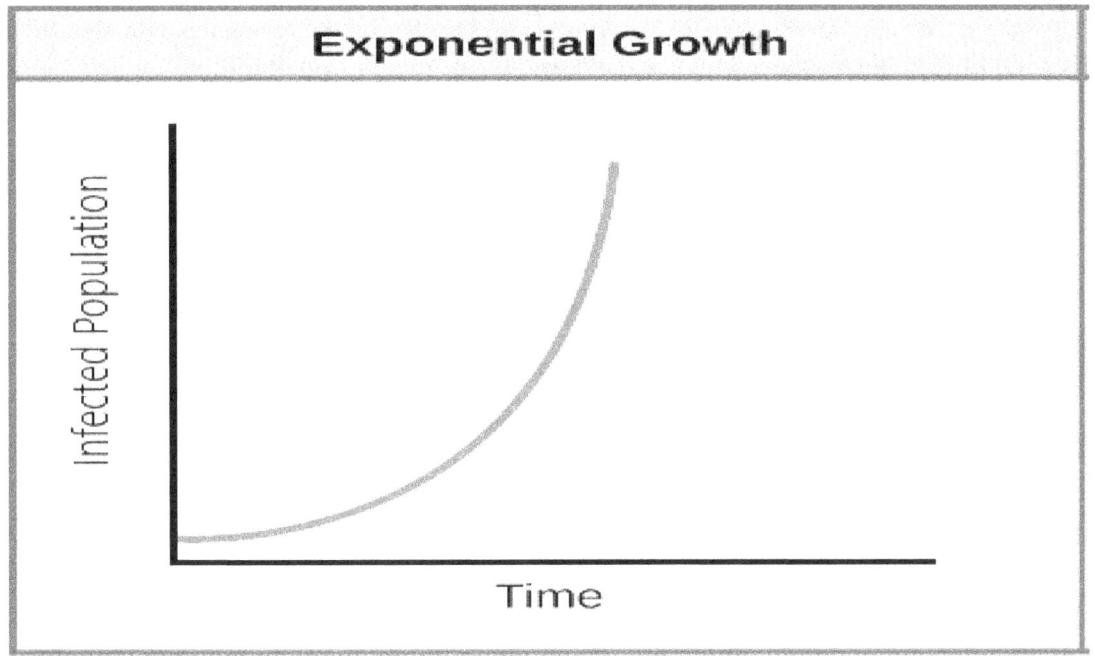

Assumptions: The rate of change of infections is proportional to the number already infected.

Limitations: The more there are that have it, the less there are that can catch it. That means the growth rate will slow down. The exponential models only works well in the early stages of a disease. Behavior can also change the growth rate.

Equation: $I(t) = I_0 e^{r_e t}$

I_0 = number initially infected with virus

r_e = growth rate (constant, unless behavior changes)

t = time

I(t) = number infected at time t

Two data points allow us to determine the parameters of the growth function, and then to predict the future trajectory of the virus. In practice, we would use more data points to filter out the noise in the data, and either do an exponential regression, or use logs and do a linear regression.

Example

> When t=0 days there are 555 known cases of a virus, and when t=3 days there are 1,400 cases. Set up the model $I(t)=I_0 e^{kt}$, using 4 significant figures for k.

Solution

> First substitute the initial population:
>
> $I(t)=I_0 e^{rt}$
>
> $I_0=555$
>
> $I(t)=555e^{rt}$
>
> Now, solve for k using the second data point.
>
> $I(t)= 555e^{rt}$
>
> $I=1,400$ when $t = 3$ days
>
> $1,400 = 555e^{r \cdot 3}$
>
> $$\frac{1,400}{555} = \frac{\cancel{555}e^{r \cdot 3}}{\cancel{555}}$$
>
> $$\frac{1,400}{555} = e^{r \cdot 3}$$
>
> $\ln(1,400/555) = \ln(e^{r \cdot 3})$

$$\ln(1{,}400/555) = 3r_e$$

$$\frac{\ln(1{,}400/555)}{3} = \frac{3r_e}{3}$$

$$r_e = \frac{\ln(1{,}400/555)}{3} = 0.3084$$

Substitute r back into the model,

$$I(t) = 555e^{0.3084t}$$

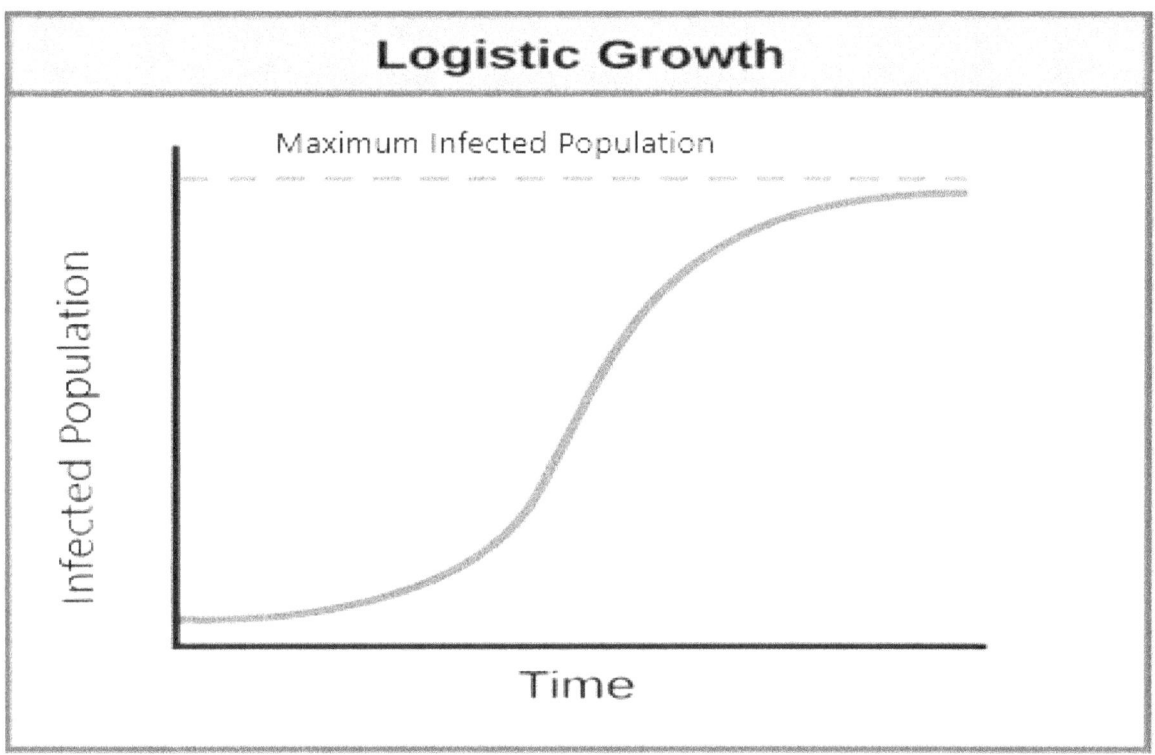

Assumptions: The rate of change of infections is proportional to the number of possible interactions of those with infected and those not infected.

Limitations: This model assumes that once you get a virus, you may infect another person indefinitely. For most strains of the flu the period of transmission is one day before symptoms, until 5-7 days after.

Equation: $$I(t) = \frac{K}{1+\frac{K-I_0}{I_0}e^{-r_1 t}}$$

$I(t)$ = population infected at time t

K = maximum population with virus (3%-11% of population with most flues)

I_0 = initial infected population when t = 0

r_1 = logistic growth rate

t = time (usually days)

To completely determine a logistic growth equation, we need (1) the initial population with the flu, I_0, (2) the number infected at some time after t = 0 days, and (3) the maximum infectible population, K.

Example

Let's assume an initial infected population of 555 (at t=0 days), and a population of 1,400 when t=3 days, and a maximum infected population of 3,000,000 (which is an educated guess). Set up the model

$$I(t) = \frac{K}{1+\frac{K-I_0}{I_0}e^{-r_1 t}}$$

Solution

First, substitute I_0 and K into logistic model.

$$I(t) = \frac{K}{1+\frac{K-I_0}{I_0}e^{-r_1 t}}$$

I_0=555, K=3,000,000

$$I(t) = \frac{3{,}000{,}000}{1 + \frac{3{,}000{,}000 - 555}{555}e^{-r_1 t}} = \frac{3{,}000{,}000}{1 + 5{,}404.405 e^{-r_1 t}}$$

Now find the growth rate, r.

$$I(t) = \frac{3{,}000{,}000}{1 + 5{,}404.405 e^{-r_1 t}}$$

I=1,400 when t=3 days

$$1{,}400 = \frac{3{,}000{,}000}{1 + 5{,}404.405 e^{-r_1 3}}$$

$$1{,}400(1 + 5{,}404.405 e^{-r_1 3}) = 3{,}000{,}000$$

$$1{,}400 + 7{,}566{,}167 e^{-r_1 3} = 3{,}000{,}000$$

$e^{-r_1 3} = 0.3963169$ (skipped a few steps here)

$$\ln(e^{-3r_1}) = \ln(0.3963169)$$

$$-3r_1 = \ln(0.3963169)$$

$$r_1 = \frac{\ln(0.3963169)}{-3} = 0.3085$$

Now substitute back into model,

$$I(t) = \frac{3{,}000{,}000}{1 + 5{,}404.405 e^{-0.3085 t}}$$

Problems

1) Letting $I_0=555$ when t=0 days, and I=2,900 when t=5 days. Set up the exponential model $I(t)=I_0 e^{rt}$. How does this compare to the example above? It should be very close.

2) Letting $I_0=555$ when t=0 days and I=2,900 when t=5 days, set up the logistic model $I(t) = \dfrac{K}{1+\frac{K-I_0}{I_0}e^{-r_1 t}}$. Use K=3,000,000. How does this compare to the example above? It should be very close.

Notes

Chapter 2: Making Predictions

The whole point of constructing a model is to make predictions. Once we have completely determined the parameters of a model – I_0, K and r_l in the logistic case – we can make predictions indefinitely far into the future.

Of course, the farther we try to look into the future, the more likely it is that the model will fail. That's why we continually update our model.

Predictions With Our Exponential Model

From Part I, our exponential model is,

$$I=555e^{0.3084t}$$

Here are a couple predictions we can make.

Example

How long will it take for the infected population to reach 10,000,000,000?

Solution

Our model is

$$I=555e^{0.3084t}$$

I=10,000,000,000

Substituting into our model,

$$10,000,000,000=555e^{0.3084t}$$

Solving for t,

$$\frac{10{,}000{,}000{,}000}{555} = \frac{555 e^{0.3084t}}{555}$$

$$18{,}018{,}018 = e^{0.3084t}$$

$$\ln(18{,}018{,}018) = \ln(e^{0.3084t})$$

$$\ln(18{,}018{,}018) = 0.3084t$$

$$t = \ln(18{,}018{,}018)/0.3084 = 54.17 \text{ days}$$

What is wrong with this prediction? Well, 10 billion people exceeds the world population of 7.8 billion!

Example

Use the exponential model to predict the number infected after 10 days.

Solution

Our model is,

$$I = 555 e^{0.3084t}$$

Substituting t=10 days,

$$I = 555 e^{0.3084(10)}$$

$$I = 12{,}124$$

Predictions With Our Logistic Model

From Part I our logistic model is,

$$I(t) = \frac{3{,}000{,}000}{1+5{,}404.405e^{-0.3085t}}$$

With this model, the population will never exceed 3,000,000, so it is pointless to ask when we will reach 10 billion. Looking for a solution will lead to an imaginary answer. We can, however, make some useful predictions.

Problem

What will the infected population be after 10 days with our logistic model?

Solution

Our model is,

$$I(t) = \frac{3{,}000{,}000}{1+5{,}404.405e^{-0.3085t}}$$

Substituting t=10 days,

$$I(10) = \frac{3{,}000{,}000}{1+5{,}404.405e^{-0.3084(10)}} = 12{,}090$$

Which is very close to the 12,124 predicted by the exponential model.

Problem

When can we expect to reach 1,500,000 infected?

Solution

Our model is,

$$I(t) = \frac{3{,}000{,}000}{1+5{,}404.405e^{-0.3085t}}$$

Substituting I=1,500,000,

$$1{,}500{,}000 = \frac{3{,}000{,}000}{1+5{,}404.405e^{-0.3085t}}$$

Solving for t,

$1{,}500{,}000(1 + 5{,}404.405e^{-0.3085t})=3{,}000{,}000$

$1 + 5{,}404.405e^{-0.3085t}=2$ (divided both sides by 1,500,000)

$5{,}404.405e^{-0.3085t}=1$

$e^{-0.3085t}=1/5{,}404.405$

$e^{-0.3085t}=0.000185034$

$\ln(e^{-0.3085t})=\ln(0.000185034)$

-0.3085t=ln(0.00185034)

t=ln(0.000185034)/(-0.3085)=27.86 days, or 28 days rounded

Problems

1) Given our exponential model $I=555e^{0.3084t}$,

a) predict the number of infections when t=20 days.

b) predict how long it will take for there to be 10,000 infections.

2) Given our logistic model $I(t) = \dfrac{3{,}000{,}000}{1+5{,}404.405e^{-0.3085t}}$

a) predict the number of infections when t=20 days.

b) predict how long it will take for there to be 10,000 infections.

3) Suppose we practice social distancing and our model changes to

$I(t) = \dfrac{3{,}000{,}000}{1+5{,}404.405e^{-0.150t}}$. Predict how long it will take for there to be 10,000 infections.

Notes

Chapter 3: What Mechanism Produces the Logistic S-Curve?

To use a simple scenario, suppose we have 20 people in a small town, and 1 has the flue. That means 19 don't. The flu can only spread when the 1 person with the flu interacts with any of the 19 without the flu. So, there are 1x19=19 possible interactions. Below we summarize this for all other scenarios.

Number With Flu, I	Number Without Flu, K-I	Possible Interactions, I(K-I)	
1	19	1x19=19	
2	18	2x18=36	
3	17	3x17=51	
4	16	4x16=64	
5	15	5x15=75	Superintendents close school here
6	14	6x14=84	
7	13	7x13=91	
8	12	8x12=96	
9	11	9x11=88	
10	10	10x10=100	Maximum rate of increase
11	9	11x9=88	
12	8	12x8=86	
13	7	13x7=81	
14	6	14x6=84	
15	5	15x5=75	
16	4	16x4=64	
17	3	17x3=51	Slowing down because not too many remain to be infected.
18	2	18x2=36	
19	1	19x1=19	
20	0	20x0=0	Good news, no new cases! Bad news, everyone is sick!

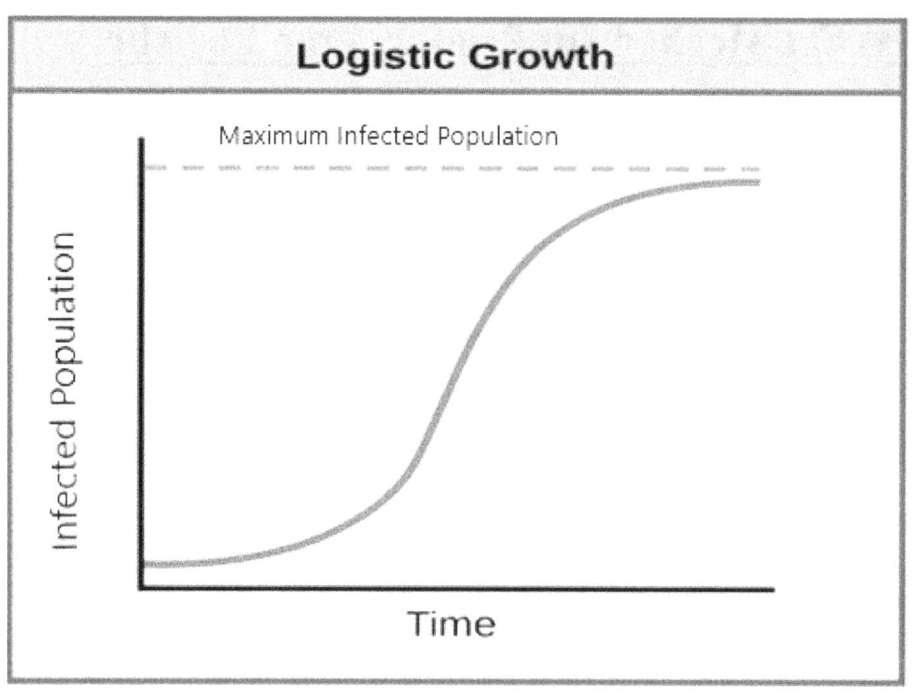

↑

Maximum Rate of Increase Here

New Infections Per Unit Time $=c(I)(K-I)$ <- <u>Differential Equation</u>

In a class called <u>Differential Equations</u>, you will use the above to produce the logistic equation we introduced in Part I. This is called <u>solving a differential equation.</u>

There are several things to note about the constant of proportionality, c:

1) Increasing c will accelerate the spread of a virus and decreasing c will decelerate the spread of the virus.

2) Primary factors for determining c are population density, virility of the disease and social habits.

3) If c and K are constant throughout the spread of the virus, the graph will be symmetric. If either vary, we will have an asymmetric graph.

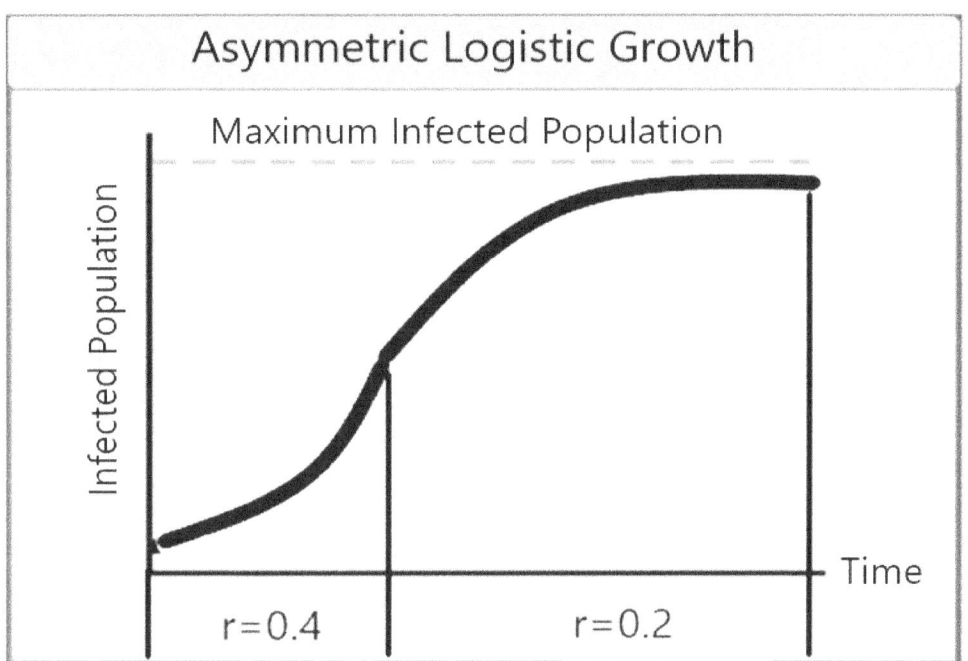

4) $cK = r_1$, where r_1 is the logistic growth rate from the previous two sections. Don't think too deeply here. You won't understand this relation until you take Differential Equations.

Problem

In the above example, let $r_1 = 0.20$. Using K=20, a) find c, and b) find the growth rates for each row in the table above.

a) $cK = r_1$

c20=0.20

c=0.20/20=0.01

b)

Number With Flu, I	Number Without Flu, K-I	Possible Interactions, I(K-I)	cI(K-I) infections/day
1	19	1x19=19	0.01x19=0.19
2	18	2x18=36	0.01x36=0.36
3	17	3x17=51	0.01x51=0.51
4	16	4x16=64	0.01x64=0.64
5	15	5x15=75	0.01x75=0.75
6	14	6x14=84	0.01x84=0.84
7	13	7x13=91	0.01x91=0.91
8	12	8x12=96	0.01x96=0.96
9	11	9x11=99	0.01x99=0.99
10	10	10x10=100	0.01x100=1.00
11	9	11x9=99	0.01x88=0.99
12	8	12x8=96	0.01x96=0.96
13	7	13x7=91	0.01x91=0.91
14	6	14x6=84	0.01x84=0.84
15	5	15x5=75	0.01x75=0.75
16	4	16x4=64	0.01x64=0.64
17	3	17x3=51	0.01x51=0.51
18	2	18x2=36	0.01x36=0.36
19	1	19x1=19	0.01x19=0.19
20	0	20x0=0	0.01x0=0.00

Just one more thing. The last column has a unit of Infections/Day. If we take the reciprocal of that column, the units become Days/Infection. That is, we will know the approximate number of days each additional infection takes.

Lets do just that.

Number With Flu, I	Number Without Flu, K-I	Possible Interactions, I(K-I)	cI(K-I) Infections/Day	Days/Infection
1	19	1x19=19	0.01x19=0.19	1/0.19=5.3
2	18	2x18=36	0.01x36=0.36	1/0.36=2.8
3	17	3x17=51	0.01x51=0.51	1/0.51=2.0
4	16	4x16=64	0.01x64=0.64	1/0.64=1.6
5	15	5x15=75	0.01x75=0.75	1/0.75=1.3
6	14	6x14=84	0.01x84=0.84	1/0.84=1.2
7	13	7x13=91	0.01x91=0.91	1/0.91=1.1
8	12	8x12=96	0.01x96=0.96	1/0.96=1.0
9	11	9x11=99	0.01x99=0.99	1/0.99=1.0
10	10	10x10=100	0.01x100=1.00	1/1.00=1.0
11	9	11x9=99	0.01x88=0.99	1/0.99=1.0
12	8	12x8=96	0.01x96=0.96	1/0.96=1.0
13	7	13x7=91	0.01x91=0.91	1/0.91=1.1
14	6	14x6=84	0.01x84=0.84	1/0.84=1.2
15	5	15x5=75	0.01x75=0.75	1/0.75=1.3
16	4	16x4=64	0.01x64=0.64	1/0.64=1.6
17	3	17x3=51	0.01x51=0.51	1/0.51=2.0
18	2	18x2=36	0.01x36=0.36	1/0.36=2.8
19	1	19x1=19	0.01x19=0.19	1/0.19=5.3
20	0	20x0=0	0.01x0=0.00	

Total Time= 35.6 days

Just one more column, and that is Cumulative Time

Number With Flu, I	Number Without Flu, K-I	Possible Interactions, I(K-I)	cI(K-I) infections/ day	Days/ infection	Cumulative Time (days)
1	19	1x19=19	0.19	5.3	0.0
2	18	2x18=36	0.36	2.8	5.3
3	17	3x17=51	0.51	2.0	8.1
4	16	4x16=64	0.64	1.6	10.1
5	15	5x15=75	0.75	1.3	11.7
6	14	6x14=84	0.84	1.2	13.0
7	13	7x13=91	0.91	1.1	14.2
8	12	8x12=96	0.96	1.0	15.3
9	11	9x11=99	0.99	1.0	16.3
10	10	10x10=100	1.00	1.0	17.3
11	9	11x9=99	0.99	1.0	18.3
12	8	12x8=96	0.96	1.0	19.3
13	7	13x7=91	0.91	1.1	20.3
14	6	14x6=84	0.84	1.2	21.4
15	5	15x5=75	0.75	1.3	22.6
16	4	16x4=64	0.64	1.6	24.9
17	3	17x3=51	0.51	2.0	25.5
18	2	18x2=36	0.36	2.8	27.5
19	1	19x1=19	0.19	5.3	30.3
20	0	20x0=0	0.00		35.6

This data can be used to produce a population vs. time graph, and we have effectively produced a numerical approximation to a solution of an autonomous, logistic differential equation!

Below is a graph of both the above discrete solution, and also the continuous solution $I(t) = \frac{20}{1+19e^{-0.20t}}$.

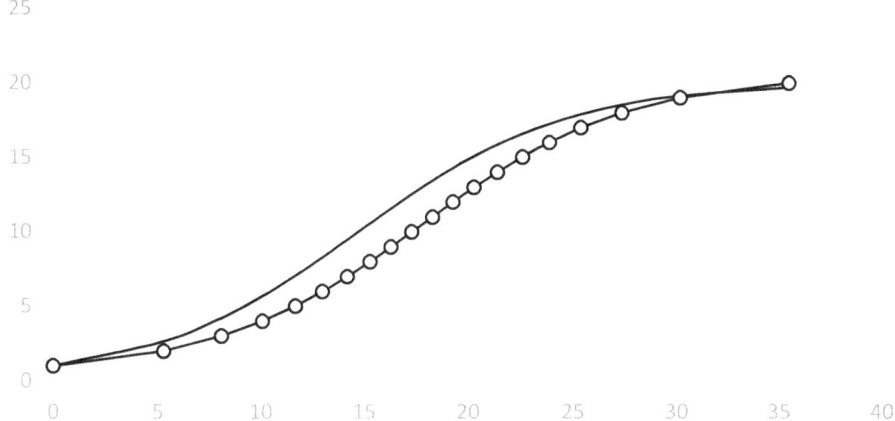

Problems

1) We have a susceptible population of K=15, initially one is infected, and we also have $r_1=0.1$.

a) Use $cK=r_1$ to find c.

b) Fill in the following table and find the total time for full infection.

Number With Flu, I	Number Without Flu, K-I	Possible Interactions, I(K-I)	cI(K-I) Infections/ Day	Days/ Infection	Cumulative Time (days)
1					
2					
3					
4					
5					
6					
7					
8					
9					
10					
11					
12					
13					
14					
15					

Total Time=

2) Returning to our case with K=20 infectible people, with one infected, suppose $r_1=0.2$ for the first 10 infections, and then $r_1=0.1$ for the last 10.

a) Use $cK=r_1$ to find c for the first 10 infections, and then c for the last 10. Use K=20 in both cases.

b) Fill in the chart below.

Number With Flu, I	Number Without Flu, K-I	Possible Interactions, I(K-I)	cI(K-I) Infections/ Day	Days/ Infection	Cumulative Time
1					
2					
3					
4					
5					
6					
7					
8					
9					
10					
11					
12					
13					
14					

15					
16					
17					
18					
19					
20					

Total Time=

c) Explain why this is an <u>asymmetric</u> S-curve.

Notes

Chapter 4: The Apex and S-Curve Basics

In the graph below we progress from 1 infection on day 0 to 41 infections on day 10. From this graph we can make a table of total cases and new cases on a daily basis.

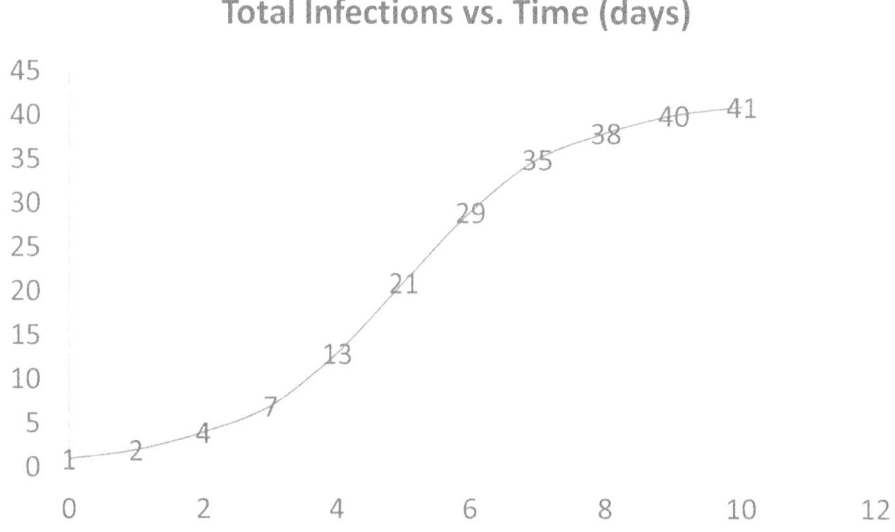

Day	Total Infections	New Infections
0	1	-
1	2	2-1=1
2	4	4-2=2
3	7	7-4=3
4	13	13-7=6
5	21	21-13=8
6	29	29-21=8
7	35	35-29=6
8	38	38-35=3
9	40	40-38=2
10	41	41-40=1

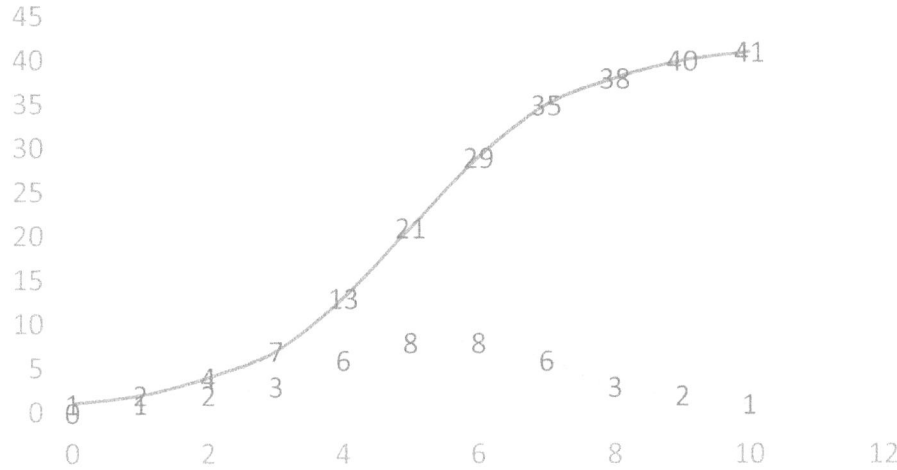

The bottom curve is the New Infections vs. Time. The apex of the curve is at t = 5.5 days, which is where the growth rate is the greatest. After t=5.5 days the new cases continue to grow, only at a slower and slower and slower rate.

Problem

1) Below is a plot of Infections vs. Time in days.

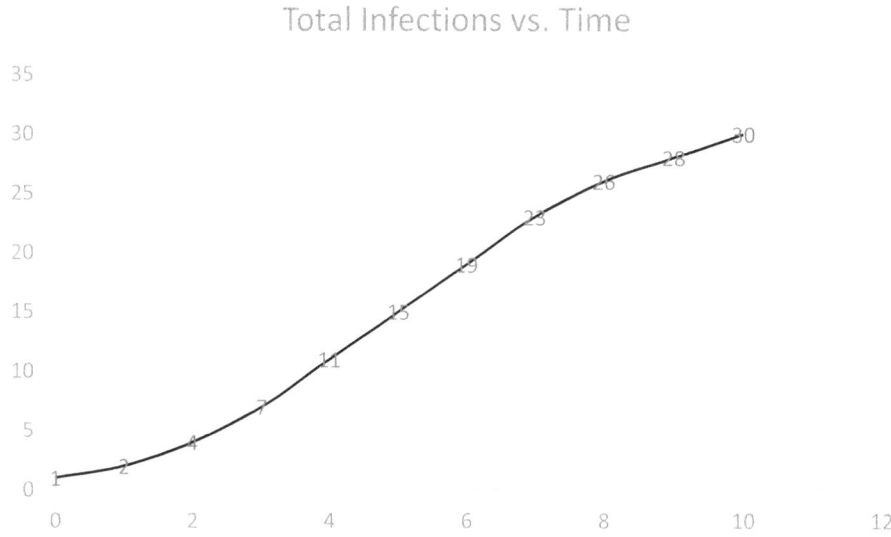

Use the graph to fill in the chart below.

Day	Total Infections	New Infections
0		
1		
2		
3		
4		
5		
6		
7		
8		
9		
10		

Use the above chart to plot New Infections vs. Time below:

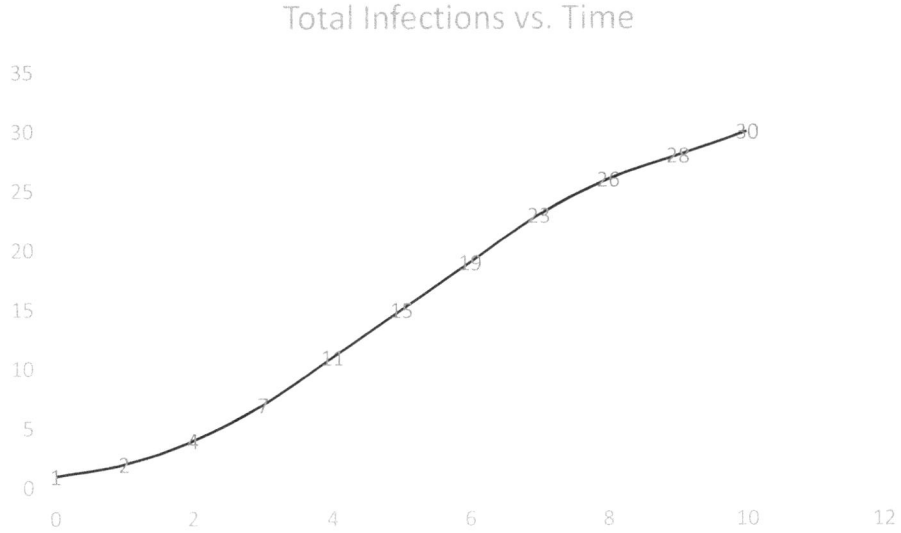

Notes

Chapter 5: Estimating the Logistic Growth Rate When $I_0 \ll K$

In Part I we looked at both an exponential and logistic growth model:
$$I(t) = 555e^{0.3084t}$$
and,
$$I(t) = \frac{3{,}000{,}000}{1 + 5{,}404.405e^{-0.3085t}}$$

It was no mistake that the growth rates are 0.3084 and 0.3085. These two parameters are nearly equal (equal to 3 significant figures, here) as long as the initial infected population is very small relative the maximum infected population. "Very small relative to" is s somewhat vague term, but useful in college-level physic and chemistry, so it is worth explaining here.

In this particular example we have $I_0=555$, and $K=3{,}000{,}000$. Had the initial population been 1,500,000, there would have been a significant difference in growth rates. The next example illustrates this.

Example

Let $I_0=1{,}500{,}000$, $I=1{,}600{,}000$ when $t=3$ days, and $K=3{,}000{,}000$.
Find (a) the exponential model $I(t) = I_0 e^{r_e t}$, and b) the logistic model
$$I(t) = \frac{K}{1 + \frac{K-I_0}{I_0} e^{-r_l t}}.$$

Solution

a) First substitute the initial population:

$$I(t) = I_0 e^{r_e t}$$

$I_0 = 1,500,000$

$$I(t) = 1,500,000 e^{r_e t}$$

Now, solve for r_e using the second data point.

$$I(t) = 1,500,000 e^{r_e t}$$

$I = 1,600,000$ when $t = 3$ days.

$$1,600,000 = 1,500,000 e^{r_e 3}$$

$$\frac{1,600,000}{1,500,000} = \frac{\cancel{1,500,000} e^{r_e 3}}{\cancel{1,500,000}}$$

$$\frac{1,600,000}{1,500,000} = e^{r_e 3}$$

$$\ln(16/15) = \ln(e^{r_e 3})$$

$$\ln(16/15) = 3 r_e$$

$$\frac{\ln(16/15)}{3} = \frac{\cancel{3} r_e}{\cancel{3}}$$

$$r_e = \frac{\ln(16/15)}{3} = 0.02151$$

Substitute r back into the model,

$$I(t) = 1,500,000 e^{0.02151 t}$$

b) $I(t) = \dfrac{K}{1+\frac{K-I_0}{I_0}e^{-r_1 t}}$

$$I_0 = 1{,}500{,}000, \quad K = 3{,}000{,}000$$

$$I(t) = \dfrac{3{,}000{,}000}{1 + \dfrac{3{,}000{,}000 - 1{,}500{,}000}{1{,}500{,}000} e^{-r_1 t}} = \dfrac{3{,}000{,}000}{1 + 1e^{-r_1 t}}$$

Now find the growth rate, r.

$$I(t) = \dfrac{3{,}000{,}000}{1 + e^{-r_1 t}}$$

I=1,600,000 when t=3 days

$$1{,}600{,}000 = \dfrac{3{,}000{,}000}{1 + e^{-r_1 3}}$$

$1{,}600{,}000(1 + e^{-r_1 3}) = 3{,}000{,}000$

$1{,}600{,}00 + 1{,}600{,}000 e^{-r_1 3} = 3{,}000{,}000$

$e^{-r_1 3} = 14/16$ (skipped a few steps here)

$\ln(e^{-r_1 3}) = \ln(14/16)$

$-3r_1 = \ln(14/16)$

$r_1 = \dfrac{\ln(14/16)}{-3} = 0.04451$

Now substitute back into model,

$$I(t) = \dfrac{3{,}000{,}000}{1 + e^{-0.04451 t}}$$

With the exponential model, we have $r_e = 0.02151$, whereas with the logistic model we have $r_1 = 0.04451$. The logistic growth rate is more than the exponential growth rate. Why? Because the initial population of 1,500,000 is a sizable fraction of K=3,000,000. We wouldn't say that 1,500,000<<3,000,000.

Why is it important to know when these two growth rates agree, approximately? There are two reasons:

(1) In the early stages of virus spread we may not know the what K is, or even a crude approximation for K. So, we can use the exponential model to approximate r_e, use this as an approximation for r_l, and then use more data points to estimate K. That would completely determine out model.

(2) Models need continual readjustments. In the case of virus spread the government and health officials induce behavior changes, and the reality of death and disease cuts very deep in peoples' psyche as well. Doing a midstream adjustment would require different analysis that just calculating an exponential growth rate.

Here is a good rule of thumb: x<<y if x is less than 1% of y.

Example

> Given the exponential model I(t)=$75e^{0.4356t}$, and K=150,000, set up the logistic model.

Solution

> I(t)=$75e^{0.4356t}$
>
> I_0=75, r_e=0.4356
>
> $$I(t) = \frac{K}{1+\frac{K-I_0}{I_0}e^{-r_l t}}$$

Let $I_0=75$, $K=150{,}000$, $r_1=0.436$ (one less significant figure)

$$I(t) = \frac{150{,}000}{1+\frac{150{,}000-75}{75}e^{-0.436t}} = \frac{150{,}000}{1+1{,}999e^{-0.436t}}$$

Problems

1) For each of the following, use r_e to estimate r_l, find I_0 and use given K to set up a logistic model. When estimating r_l, drop 1 significant figure.

Exponential Model	r_e	Estimated r_l	I_0	K	$I(t) = \dfrac{K}{1 + \dfrac{K - I_0}{I_0} e^{-r_l t}}$
$I(t)=12e^{0.1324t}$				5,000	
$I(t)=125e^{0.0324t}$				10,000	
$I(t)=150e^{0.1382t}$				25,000	

2) Circle TRUE or FALSE. Use the rule given in the last sentence of this section.

a) $2 \ll 300$ TRUE   FALSE   b) $1 \ll 1000$ TRUE   FALSE

c) $3.2 \ll 59$ TRUE   FALSE   d) $22 \ll 23$ TRUE   FALSE

Notes

Chapter 6: Doubling Time and the Growth Rate

A characteristic of an exponential growth function is that the population will double at equal time intervals.

For example,

$$I(t)=20e^{0.3466t}$$

will double every 2 days. So, we would predict the population to grow from 20 to 40, 80, 160, 320 in 2, 4, 6 and 8 days, respectively.

To verify,

$$I(8)=20e^{0.3466(8)}=320.0676.$$

This gives us a useful way to quickly build an exponential model, with the following formula:

$$r_e = \frac{\ln(2)}{\text{time-to-double}}$$

Problem

Initially a virus is doubling every 2.3 days, with an initial infected population of 555. Build (a) the model $I(t) = I_0 e^{r_e t}$ and then (b) the model $I(t) = \frac{K}{1+\frac{K-I_0}{I_0}e^{-r_1 t}}$.

a) Note that

$$r_e = \frac{\ln(2)}{\text{time-to-double}}$$

time-to-double= 2.3 days

$$r_e = \frac{\ln(2)}{2.3} = 0.3014 \text{ (as opposed to 0.3084)}$$

Now build the model

$$I(t) = I_0 e^{r_e t}$$

$I_0 = 555$, $r_e = 0.3014$

$$I(t) = 555 e^{0.3014 t}$$

b) Since the 555<<3,000,000, we will use a logistic growth rate of $r_l = 0.301$ (1 less sig. fig.)

$$I(t) = \frac{K}{1 + \frac{K - I_0}{I_0} e^{-r_l t}}$$

$K = 3{,}000{,}000$, $r_l = 0.301$, $I_0 = 555$

$$I(t) = \frac{3{,}000{,}000}{1 + \frac{3{,}000{,}000 - 555}{555} e^{-0.301 t}} = \frac{3{,}000{,}000}{1 + 5404.405 e^{-0.301 t}}$$

This doubling principle is useful information because we can spot when the spread of a virus is transitioning away from exponential growth – either due to following logistic growth, or due to social distancing and other measures.

Below are infections for two time-segments for a virus:

3/21	**555**	
3/22	654	
3/23	941	
3/24	1,400	1/21 figure doubled after about 2.3 days

5/21	**337,000**	
5/22	378,200	
5/23	418,000	
5/34	467,700	
5/35	529,600	
5/26	593,300	
5/27	660,700	5/21 figure nearly doubled after 6 days
5/28	720,100	
5/29		

Within two months we moved well beyond the initial exponential growth period.

Problems

1) Fill in the following table:

Initial Infected Population	Days For Population To Double	$r_e = \dfrac{\ln(2)}{\text{time-to-double}}$	Model, $I(t) = I_0 e^{r_e t}$
100	5	0.1386	$I(t) = 100 e^{0.1386 t}$
200	4		
1,000	10		
1	7		

2) Fred has $10 in a drawer and saved $2 per day. After 5 days Fred has 20, so his savings doubled. $r_e = \ln(2)/5 = 0.1386$, and $p(t) = 10 e^{0.1386 t}$ will tell you how much money Fred has after t days. So, for instance, $P(90) = 10 e^{0.1386(90)} = \$2{,}614{,}504.34$. Fred will be rich in 90 days! Explain where the fault lies in our reasoning.

3) The following is a table of infections. Use the initial doubling time to set up an exponential model.

Time (days)	Infections
0	25
1	30
2	36
3	41
4	47
5	50
6	56

$I_0 =$ _____ time-to-double = _____ $r_e =$ _____ $I(t) =$ _____

Notes

Chapter 7: What About R_0 and R_e?

R_0 is the most often talked about statistic during any particular outbreak – and also the most understood.

R_0 by definition is the average number of infections caused by one infected person, in an ideal case where the population has no prior immunity. Prior immunity could come from vaccines for the current virus, secondary immunity from prior vaccines, or secondary immunity from exposure to other viruses.

Calculating R_0 is always done indirectly with complex models. The assumption is that we are starting with a pure population, uncorrupted by any vaccines and any prior exposure to viruses. Such populations are almost non-existent.

According to the Center for Disease Control, R_0 for measles is 12-18. The variation accounts for population densities and social practices. For example, in a suburban center where children are often in daycare, R_0 will be closer to 18; whereas, out on the prairie where children a more likely than not to grow up on a farm, R_0 is closer to 12.

Unfortunately, R_0 is almost useless for modeling purposes because population immunity is not taken into account. Even though measles had an R_0 many times greater than most flu viruses, it will not spread because of widespread measles vaccinations.

A more useful statistic is R_e, which by definition is the average number of people that each infectious person will infect in a current population. This can be put directly into a model.

R_e will change through the trajectory of a virus, simply because of herd immunity.

R_0 = basic reproduction number (for totally susceptible population)
R_e = effective reproduction number (for actual population)
$R_e = xR_0$, x = percentage susceptible to a particular virus

Consider the following example:

> An aboriginal tribe with population 1000 is exposed to a virus with $R_0=3$. The tribe has no prior immunity through exposure to any virus nor through any vaccines. Over 20 days the whole tribe gets the virus.

Below is a logistic curve describing the trajectory of the virus, and what happens to R_e as the disease progresses.

$R_0=3$	$R_0=3$	$R_0=3$
$R_e=3$	$R_e=1.5$	$R_e=0$
Everyone is susceptible	Half the population is susceptible	No one is susceptible

Generally,

  If $R_e>1$, the number infected will grow

  If $R_e=1$, we have reached equilibrium. The number infected will remain constant.

  If $R_e<1$, the number infected will decrease.

If $R_e>1$ we have several choices:

(1) Wait until enough people develop immunity, at which time $R_e<1$, and the virus dies out.

(2) Quarantine known carriers and anyone exposed to such carriers to extinguish the virus. This is effective in the early stages of a virus, but impractical soon after.

(3) Social distancing.

Whatever is put into practice, if $R_e>1$, we need to quarantine or immunize greater than $(1-1/R_e)$ of the population for extinguishing to commence.

Example

>For a flu virus let $R_e=2.3$.
>
>$1-1/R_e=1-1/2.3=0.5652=56.52\%$ need immunity to the virus.
>
>Path A: If we immunize more than 56.52% of the population, then the flu virus will die out on its own. The number of new infections will decrease.
>
>Path B: If 56.52% of the population is infected, and the infected all develop immunity – not a given for every virus – then the virus will also die out with the same final trajectory as Path A.

The vaccination path is better because, if done early, not a lot of people will have to suffer through our virus, and our healthcare system will not be stressed.

Problems

1) Fill in the following table

R_e	Fraction vaccinated for equilibrium $1-1/R_e$	Fraction without immunity, $x=1/R_e$	xR_e
1.1			
1.8			
2.1			
5.0			
15.0			

2) Fill in the following table.

Initial R_e	Percent With Herd Immunity	Percent Without Herd Immunity	New R_e
2.5	28%	1-28%=72%	0.72x2.5=1.8
5.5	80%		
7.0	92%		

3) A virus has an initial R_e value of 3.2. Find the minimum percentage that must be vaccinated to establish equilibrium.

4) Assume R_e=15 for measles. What percentage must be vaccinated to establish equilibrium?

Notes

Chapter 8: r_e and the Daily Growth Rate

In the model $I(t)=555e^{0.3081t}$, $r_e=0.3081$. r_e is the daily growth rate, compounded continuously. What does this mean?

Let's assume the rate is 0.30, for simplicity sake. For half a day the rate would be 0.15. If we compound the growth twice per day, we would have $(1+0.15)(1+0.15)=1.3225$. So, the growth rate is really 32.25% per day. If we compound four times per day, then we have $(1+0.0725)(1+0.0725)(1+0.0725)(1+0.0725)=1.3231$, and we have a daily growth rate of 32.31% We can continue this process, summarized in the following table:

Compounding	Calculation	Daily Growth Rate
Daily	$(1+0.30)-1$	0.30=30%
Twice a day	$(1+0.15)^2-1$	0.3225=3.225%
Four times per day	$(1+0.075)^4-1$	0.3231=32.31%
Ten times per day	$(1+0.03)^{10}-1$	0.3438=34.39%
One hundred times per day	$(1+0.003)^{100}-1$	0.3492=34.92%
An infinite times per day	$e^{0.30}-1$	0.3499=34.99%

Of course we can't really do this process an infinite number of times per day, but we could do it more and more times, and find a limiting value, and that would be 34.99%. This rate, 34.99%, is compounded continuously.

There is a simple way to calculate this, instead of making a table.

Let d = the daily rate of increase in infections,
$$d = e^{r_e} - 1$$

Example

Given the model $I(t) = 25e^{0.20t}$, find the initial population and the daily growth rate.

Solution

Since $I(t) = 25e^{0.20t}$,

$I_0 = 25$ infected

$r_e = 0.20$

Since $d = e^{r_e} - 1$,

$d = e^{0.20} - 1 = 0.2214 = 22.14\%$.

We can invert this relationship, giving us r in terms of d,
$$r_e = \ln(d+1)$$

Example

There are initially 50 infected, and infections are growing daily at a rate of 15%. Give the exponential model $I(t) = I_0 e^{r_e t}$.

Solution

$r_e = \ln(d+1)$

$d = 0.15$

$r_e = \ln(1+0,15) = 0.1398$

$I(t) = I_0 e^{r_e t}$

$I_0=50$, $r_e=0.1398$

$I(t) = 50e^{0.1398t}$

Problem

In the above population problem, assume we have a population of 20,000, and that everyone will get the virus. Set up the logistic model $I(t) = \dfrac{K}{1+\frac{K-I_0}{I_0}e^{-r_l t}}$.

Solution

$I(t) = \dfrac{K}{1+\frac{K-I_0}{I_0}e^{-r_l t}}$

$K=20{,}000$

$I_0=50$

$r_l=r_e=0.140$ (since $p_0 \ll K$, use one less sig. fig.)

$I(t) = \dfrac{20{,}000}{1+\frac{20{,}000-50}{50}e^{-0.140t}} = \dfrac{20{,}000}{1+399e^{-0.140t}}$

Problems

1) Fill in the following table.

Exponential Model	r_e	Daily Growth Rate, $d = e^{r_e} - 1$
$I(t) = 40e^{0.15t}$	0.15 = 15%	$e^{0.15} - 1 = 0.16 = 16\%$
$I(t) = 80e^{0.20t}$		
$I(t) = 21e^{0.25t}$		
$I(t) = 120e^{0.18t}$		
$I(t) = 60e^{0.35t}$		

2) Fill in the following table.

Daily Growth Rate, d	$r_e = \ln(d+1)$	I_0	Model, $I(t) = I_0 e^{r_e t}$
20%	$\ln(0.20+1) = 0.18$	20	$I(t) = 20e^{0.18t}$
35%		15	
0.12		45	
0.28		150	

3) A population has a daily growth rate of 18%, with 20 infected, and a total population of 2000. a) Set up the logistic model $I(t) = \dfrac{K}{1 + \frac{K - I_0}{I_0} e^{-r_1 t}}$.

b) Calculate I(1). c) Calculate 20(1+0.18).

Notes

Chapter 9: Using R_e to find r_e

The effective production number, R_e, is the number of infections caused by one infected person. R_e will vary during the spread of a virus, as a population develops herd immunity.

If we let

τ=the duration of contagious period

Then

d=R_e/τ, gives us the daily rate of the spread of the virus.

And recall that

r_e=ln(d+1) gives us the continuously compounded rate for the exponential model.

Substitution gives us,

r_e=ln(R_e/τ +1)

Problem

With a flu virus assume R_e=2.3, τ=5.5 days, I_0=555, and K=200,000,000. Set up a) the exponential model, $I(t)=I_0 e^{r_e t}$ and b) the logistic model,

$$I(t) = \frac{K}{1 + \frac{K-I_0}{I_0} e^{-r_1 t}}$$

Solution

a) $I(t) = I_0 e^{r_e t}$

$I_0 = 555$, $\tau = 5.5$ days, $R_e = 2.3$

$r_e = \ln(R_e/\tau + 1) = \ln(2.3/5.5 + 1) = \ln(1.4182) = 0.3494$

$I(t) = 555 e^{0.3494 t}$

b) $I(t) = \dfrac{K}{1 + \dfrac{K - I_0}{I_0} e^{-r_l t}}$

$K = 200{,}000{,}000$

$I_0 = 555$

$r_l = r_e = 0.349$ (since $I_0 \ll K$)

$I(t) = \dfrac{200{,}000{,}000}{1 + \dfrac{150{,}000{,}000 - 555}{555} e^{-0.349 t}} = \dfrac{200{,}000{,}000}{1 + 270{,}269.27 e^{-0.349 t}}$

Problems

1) Fill in the following table to find the daily rates.

R_e	τ days	d (infections/day)
15	3.0	
2.8	5.5	
1.2	4.0	
1.8	5.0	

2) Fill in the following table

R_e	τ days	$r_e=\ln(R_e/\tau+1)$
15	3.0	
2.8	5.5	
1.2	4.0	
1.8	5.0	

3) There are initially 12 infections in a large urban population, and $R_e=1.8$ with $\tau=7$ days. Set up an exponential model using these parameters.

Notes

Chapter 10: Social Distancing and R_e

When a major outbreak of a virus occurs and there is not enough time to develop a vaccine, a stop-gap measure is to practice social distancing. Social distancing will slow the spread of the virus down until a more permanent solution of a vaccine is available.

Exactly how social distancing works is a bit complex. Consider the following scenarios.

Scenario 1 On the nightly news it is announced that a deadly virus is expected to infect 2/3 of the population with mortality rates of 1.5%. That night a lady is seen leaving the city in a compact car packed to the roof with all of her belongings. There were various articles of women's clothing hanging out the window, flapping in the wind. The driver, who was the sole occupant, was wearing a mask and goggles. Our susceptible population was clearly now K-1.

Scenario 2 During an outbreak of a virus Bill only practices social distancing while walking down the street, but parties with friends at night in the woods behind his house. Since it is the same group of people every night, no precautions are taken. However, over half of the partiers practice no form of social distancing during the day. Bill will most likely get the virus, only it may take a little longer. His behavior may decrease the rate, r_1, but probably not the maximum infected population, K.

We will make the assumption here that social distancing will have a direct impact on the total number who will get the disease, K, which in turn will change r_1. This may be a bit of a simplification, but we need to start somewhere.

Returning to our case from section 4 with 20 infectible people, recall that,

New Infections Per Unit Time $=c(I)(K-I)$

We can derive a formula for c in terms of knowable quantities as,

$c = r_e/K = \ln(R_e/\tau + 1)/K$

And if we practice social distancing and we have maximum infected population and effective reproduction numbers of

xR_e and xK

Our formula becomes,

$c' = \ln(xR_e/\tau + 1)/(xK)$

Where,

R_e = effective reproduction number

τ = time an infected person is contagious

K = maximum infected population

x = fraction remaining infectible after social distancing

Problem

Consider a population of K=20, all infectible, with R_e=2.5, τ=5 days, and x=0.50. In other words, half are going to be removed from the infectible population.

a) Find c for the original population, and then c' for the socially distanced population. b) Find out how long it takes for total infection for each.

Solution

a) $c = \ln(R_e/\tau + 1)/K$

$R_e = 2.5$, $\tau = 5.5$ days, $K = 20$

$c = \ln(2.5/5.5 + 1)/20 = 0.0187$

$c' = \ln(xR_e/\tau + 1)/(xK)$

$R_e = 2.5$, $\tau = 5.5$, $K = 20$, $x = 0.50$

$c' = \ln(0.5(2.5)/5.5 + 1)/(0.5(20)) = 0.0205$

Number With Flu, I	Number Without Flu, K-I	Possible Interactions, I(K-I)	cI(K-I) infections/day	Days/infection
1	19	1x19=19	0.36	2.78
2	18	2x18=36	0.67	1.49
3	17	3x17=51	0.95	1.05
4	16	4x16=64	1.2	0.83
5	15	5x15=75	1.4	0.71
6	14	6x14=84	1.57	0.64
7	13	7x13=91	1.7	0.59
8	12	8x12=96	1.8	0.56
9	11	9x11=99	1.85	0.54
10	10	10x10=100	1.87	0.53
11	9	11x9=99	1.85	0.54
12	8	12x8=96	1.8	0.56
13	7	13x7=91	1.7	0.59
14	6	14x6=84	1.57	0.64
15	5	15x5=75	1.4	0.71
16	4	16x4=64	1.2	0.83
17	3	17x3=51	0.95	1.05

18	2	18x2=36	0.67	1.49
19	1	19x1=19	0.36	2.78
20	0	20x0=0	0.01x0=0.00	

Total Time=18.91 days

Number With Flu, I	Number Without Flu, K-I	Possible Interactions, I(K-I)	cI(K-I) infections/day	Days/infection
1	9	1x19=19	0.18	5.56
2	8	2x18=36	0.33	3.03
3	7	3x17=51	0.43	2.33
4	6	4x16=64	0.49	2.04
5	5	5x15=75	0.51	1.96
6	4	6x14=84	0.49	2.04
7	3	7x13=91	0.43	2.33
8	2	8x12=96	0.33	3.03
9	1	9x11=99	0.18	5.56
10	0	10x10=100	0.18	5.56

Total Time=33.34 days

Social distancing not only decreased the number of infections but stretched out the duration of the infection cycle as well.

Below is a graph of the continuous models for each.

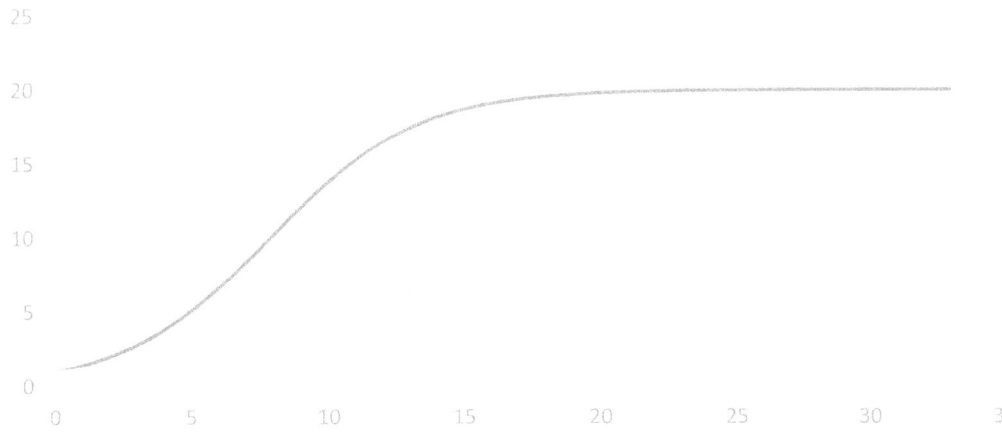

Problem

1) We have 14 people with 1 infected. $R_e=3.5$, $\tau=5$ days, and $x=0.5$ (50% removed due to social distancing).
a) Find c and c'.   b) Estimate the total time before and after social distancing.

a) c=

  c'=

b)

Number With Flu, I	Number Without Flu, K-I	Possible Interactions, I(K-I)	cI(K-I) infections/day	Days/infection
1				
2				
3				
4				
5				
6				
7				
8				
9				
10				
11				
12				
13				
14				

Total Time =

Number With Flu, I	Number Without Flu, K-I	Possible Interactions, I(K-I)	c'I(K-I) infections/day	Days/infection
1				
2				
3				
4				
5				
6				
7				

Total Time =

Notes

More Notes

Chapter 11: The SIR Model Simply Explained

The SIR model is an improvement on the simple logistic model because infected patients are removed from the contagious population once they are no longer contagious. Other than that, the SIR model is identical to the logistic model. The SIR model will still give us an S-curve, but technically it is not logistic because it does not come from the simple equation new infections per unit time = $c(I)(K-I)$.

SIR is an acronym for **S**usceptible, **I**nfected, **R**ecovered.

Susceptible – the segment of the population that can catch the virus. We assume each person can only catch the virus once.

Infected – the segment of the population infected with the virus. Infected individuals are assumed to be contagious while infected, and not contagious once recovered.

Recovered – the segment of the population that has recovered from the virus, after being infected. Recovered individuals are never susceptible nor contagious.

We assume that each person is in one of the above categories only at any given time. Gradually, everyone moves from susceptible to infected to recovered.

If a person is vaccinated, then they are on the sidelines and they are not included in the model. For instance, if a population of 1000 includes 300 vaccinated, then we will apply the model to the remaining 700.

If you have ever bowled, this is not much more complicated than keeping score.

We define the following variables:

$S(t)$=number susceptible at time t.

$I(t)$=number infected at time t.

$R(t)$=number recovered at time t.

R_e=initial effective reproduction number (how many each infected person will infect)

τ=days an infected person is contagious.

$\dfrac{R_e}{\tau(S(0)+I(0)+R(0))} \cdot S(t)I(t)$ is the New Infections for one day

$\dfrac{1}{t}I(t)$ is the New Recoveries for one day

Note: $S(0)+I(0)+R(0)$ is the total population, excluding the vaccinated.

$\dfrac{R_e}{\tau(S(0)+I(0)+R(0))} \cdot S(t)I(t)$ is really the same as $c(I)(K-I)$ from logistic growth.

$\dfrac{1}{t}I(t)$, represents the fraction of the infected that recover each day. For example, if $\tau=5$ days, then 1/5 of the infected will recover in any given day. This term is what was missing with logistic growth.

Applying the Model

Suppose we have a population of 1000, with 300 vaccinated with a 100% effective vaccine. 10 initially have the virus, with $R_e=1.8$. Also, $\tau =4$ days, so a person is contagious for 4 days. Use the discrete SIR method through 4 days.

1) Find the constant coefficients $\dfrac{R_e}{\tau(S(0)+I(0)+R(0))}$ and $\dfrac{1}{\tau}$.

$$\dfrac{R_e}{\tau(S(0)+I(0)+R(0))} = \dfrac{1.8}{4(690+10+0)} = \dfrac{1.8}{2800} = 0.000643 \qquad \dfrac{1}{\tau} = \dfrac{1}{4} = 0.25$$

2) Fill the row for day zero:

day	S(t)	I(t)	R(t)	New Infections $\dfrac{R_e}{\tau(S(0)+I(0)+R(0))} \cdot S(t)I(t)$	New Recoveries $\dfrac{1}{\tau}I(t)$
0	690	10	0	0.000643(690)(10)=4	0.25(10)=3
1					
2					
3					

3) For all days that follow:

Susceptible for current day = Susceptible from previous day

$\quad$ − New Infections from previous day

day	S(t)	I(t)	R(t)	New Infections $\dfrac{R_e}{\tau(S(0)+I(0)+R(0))} \cdot S(t)I(t)$	New Recoveries $\dfrac{1}{\tau}I(t)$
0	690	10	0	0.000643(690)(10)=4	0.25(10)=3
1	690−4=686				

Infected for current day = Infected from previous day

$\quad$ + New Infections from previous day

$\quad$ − New Recoveries from previous day

day	S(t)	I(t)	R(t)	New Infections $\dfrac{R_e}{\tau(S(0)+I(0)+R(0))} \cdot S(t)I(t)$	New Recoveries $\dfrac{1}{\tau}I(t)$
0	690	10	0	0.000643(690)(10)=4	0.25(10)=3
1	690−4=686	10+4−3=11			

Recovered for current day = Recovered from previous day

+ New Recoveries from previous day.

day	S(t)	I(t)	R(t)	New Infections $\frac{R_e}{\tau(S(0)+I(0)+R(0))} \cdot S(t)I(t)$	New Recoveries $\frac{1}{\tau}I(t)$
0	690	10	0	0.000643(690)(10)=4	0.25(10)=3
1	690-4=686	10+4-3=11	0+3=3		

…and then calculate New Infections and New Recoveries.

day	S(t)	I(t)	R(t)	New Infections $\frac{R_e}{\tau(S(0)+I(0)+R(0))} \cdot S(t)I(t)$	New Recoveries $\frac{1}{\tau}I(t)$
0	690	10	0	0.000643(690)(10)=4	0.25(10)=3
1	690-4=686	10+4-3=11	0+3=3	0.000643(686)(11)=5	0.25(11)=3

Continue this pattern .....

day	S(t)	I(t)	R(t)	New Infections $\frac{R_e}{\tau(S(0)+I(0)+R(0))} \cdot S(t)I(t)$	New Recoveries $\frac{1}{\tau}I(t)$
0	690	10	0	0.000643(690)(10)=4	0.25(10)=3
1	690-4=686	10+4-3=11	0+3=3	0.000643(686)(11)=5	0.25(11)=3
2	686-5=681	11+5-3=13	3+3=6	0.000643(681)(13)=6	0.25(13)=3
3	681-6=675	13+6-3=16	6+3=10	0.000643(675)(16)=7	0.25(16)=4

One note on rounding. You may round to the nearest person, or the nearest tenth of a person, as you please. In the long run, it really doesn't matter, unless you are dealing with a very small population.

This procedure is best done with a spreadsheet. But, doing a little by hand is worthwhile for understanding the model.

This model is usually carried out by solving a system of differential equations to produce a continuous solution. This discrete solution – i.e., by stepping though one finite interval at a time – works just as well, however, because both the continuous and the discrete are equally valid approximations of what will actually happen in the real world.

Below is a graph of a logistic approximation and also an SIR approximation for the problem above.

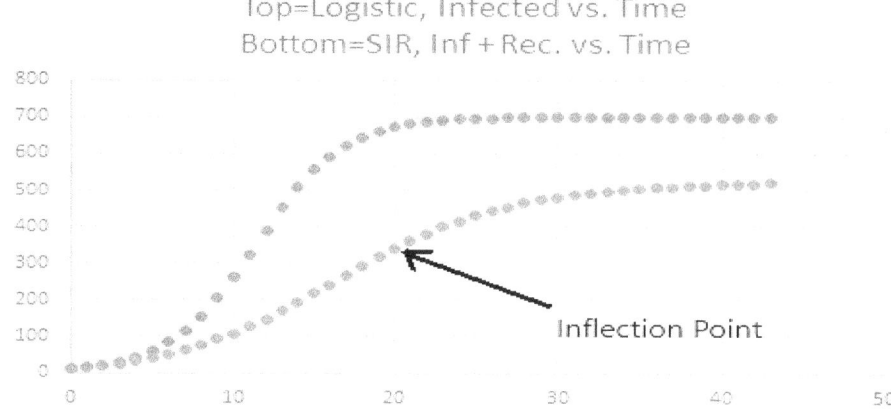

There are a couple things worth noting:
- The logistic graph tops out at 700 infections pretty quickly. That is because once infected, always infected with the logistic model, which gives us more people to infect the susceptibles.
- With the SIR model we reach an inflection point after 20 days, with 320 infections + recovered, or 380 susceptibles. It is no surprise that 380/700x1.8 is approximately 1, which is what R_e is at that point. So, the virus will enter the "dying out" phase.

One final thing is worth noting. Below is an SIR projection for measles with R_e=15 in an urban setting with a population of 1,000,000. We will assume $\tau = 7$ days. This is purely hypothetical, as we are assuming that no one is vaccinated and no one has immunity from prior exposure.

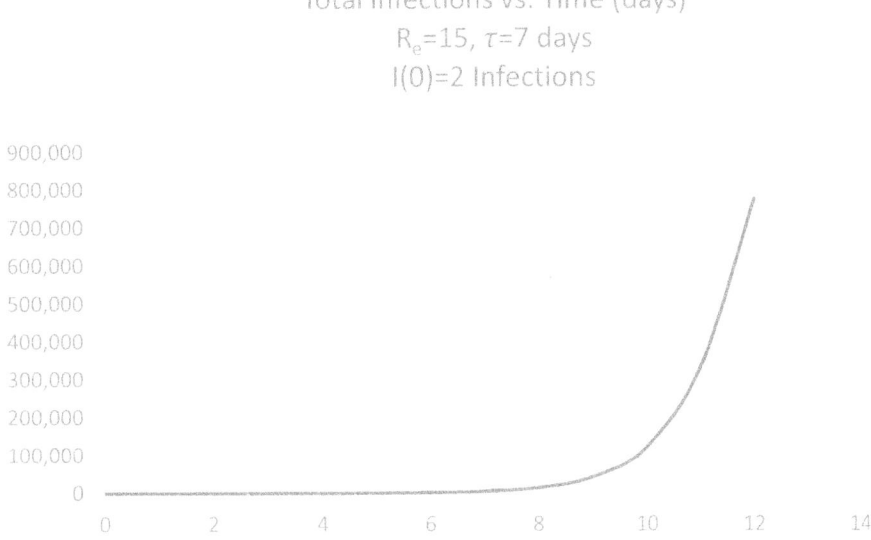

So, what can go wrong with the SIR model?

Short answer: nothing. IF our assumptions are correct, then SIR is very accurate. But,

1) If there is a significant period between exposure and being contagious, we would see a time lag, and that would slow things down a bit.

2) In the early stages of an outbreak, data collection can be sketchy. Sample sizes are often small, and adherence to protocols may vary from location to location.

3) When epidemiologists study virus spread, don't control variables in a laboratory. They collect data of for events that already happened. They are detectives reconstructing the scene of a virus.

4) Viruses mutate - sometimes for the better, and sometimes for the worse.

5) Parameters such as τ are averages and may vary from person to person, and country to country. That's why we need a lot of studies.

6) The projection assumes we have one homogeneous population. In reality we have sub populations within one country, and of course many countries that all follow their own SIR path. We may have many different S-curves, superimposed.

7) Human behavior can change the outcome, as it has with major outbreaks like the Spanish flue of 1918. These forecasts are compelling reasons why we need to change our behavior. Physicists are lucky because electrons behave as they should, in a much more predictable manner than people.

Note that 1-7 above do not, in any way, call into question the veracity of the epidemiologist or the validity of their science. All this means is we have less certainty in the beginning and more certainty with many studies, and an epidemiologist needs more studies than a mere scientist in a laboratory.

Problem

1) Suppose we have a population of 1000, and 500 vaccinated with a 100% effective vaccine. 10 initially have the virus, with R_e=2.8. Also, τ =5 days, so a person is contagious for 5 days. Use the discrete SIR method through 5 days.

$$\frac{R_e}{\tau(S(0)+I(0)+R(0))} = \qquad \frac{1}{\tau} =$$

day	S(t)	I(t)	R(t)	New Infections $\frac{R_e}{\tau(S(0)+I(0)+R(0))} \cdot S(t)I(t)$	New Recoveries $\frac{1}{\tau}I(t)$
0					
1					
2					
3					
4					
5					

Notes

More Notes

Chapter 12: Some Projections Based Upon the SIR Model

In the early stages of any virus, projections may vary, or given as a range of values. There are valid reasons for this, as we discussed in the previous section. In this section we will vary some parameters and look at the effect. The SIR model will be used and implemented with a spreadsheet.

Varying the Length of the Infectious Period

Our first projections look at varying the time an average person is contagious, τ. We will assume K=330,000,000 (approximately the population of the U.S.), R_e=2.3 and τ varies from 5.5 days to 11 days. In other words, we will lengthen τ from 5.5 days to 11 days, yet everyone still infects 2.3 people. As we may expect, things will slow down considerably.

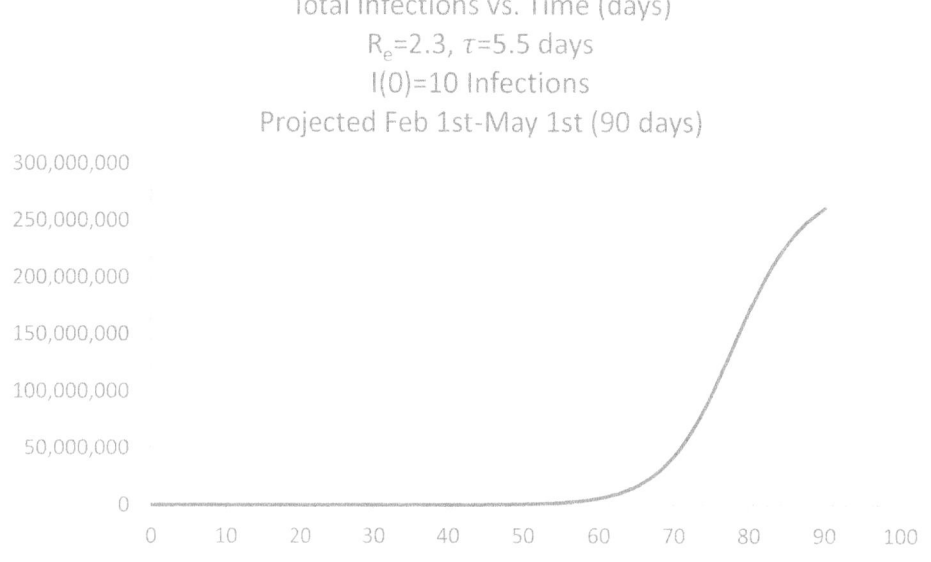

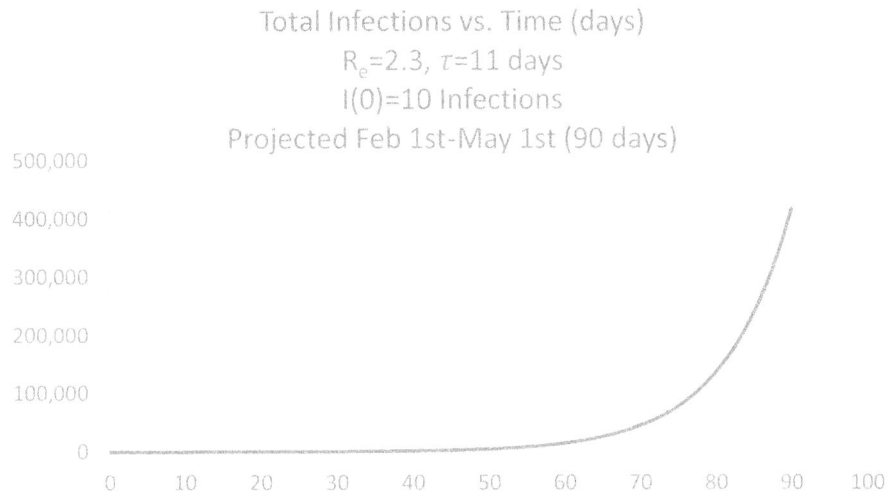

There is a significant decrease in infections through 90 days – over 250 million for τ=5.5 days, and a little over 400 thousand when τ=11 days.

Varying the Effective Reproduction Number, R_e

Our next projections look at varying R_e. We will assume τ=7 days, K=330,000,000 people, and use R_e=1.1, R_e=1.8, R_e=2.3 and R_e=5.8. As you will see, a lot turns on this single parameter.

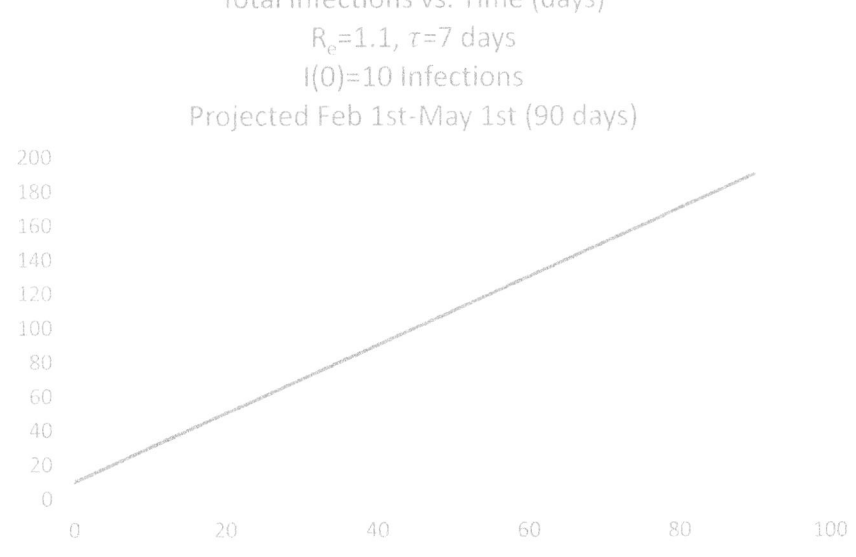

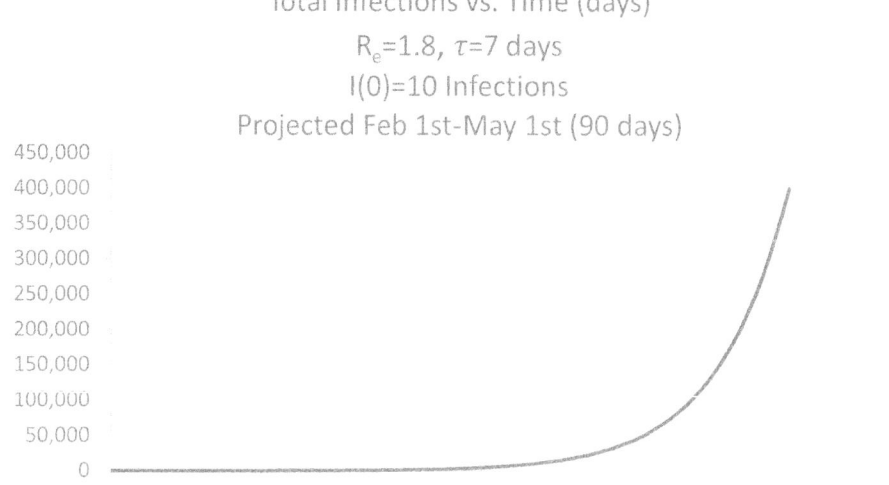

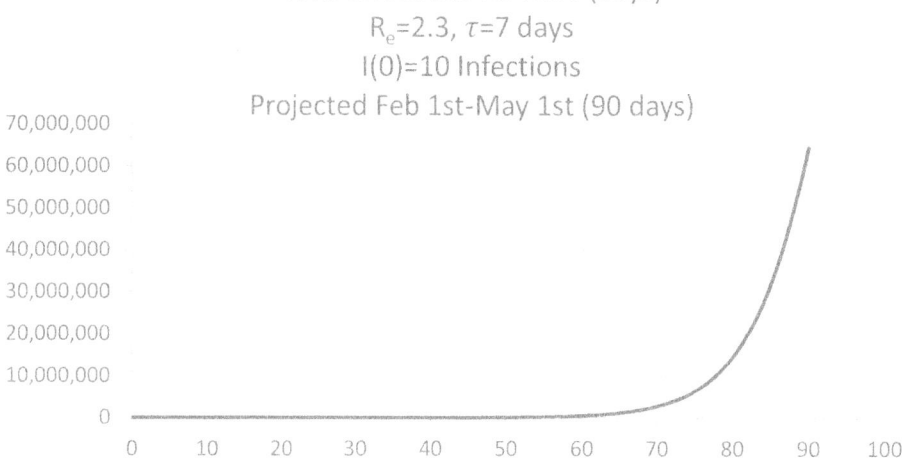

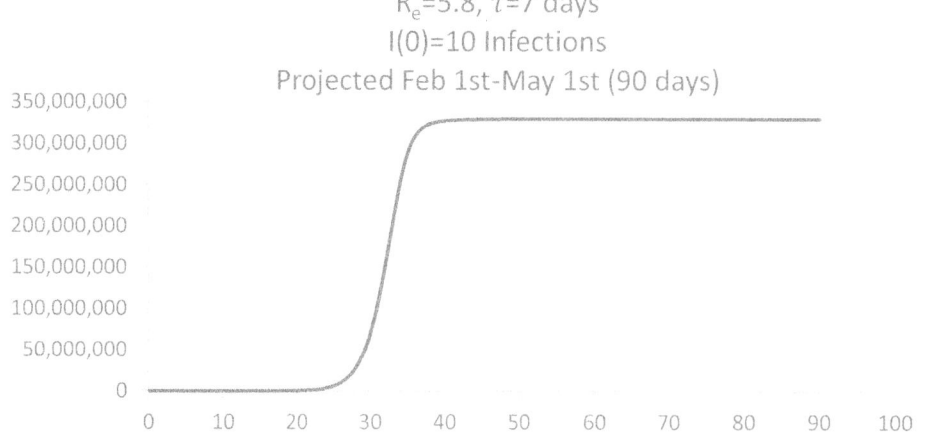

For the Spanish flu of 1918 $R_e=1.8$ was the median figure. The Swine flu of 2009 spread a little slower: $R_e=1.46$ for a first wave, and $R_e=1.48$ for a second wave.

Epidemiologists often give a range of values in their projections, lacking certainty, especially in the early stages. There is certainly here, however, but it is statistical certainty. When a very early epidemiological figure is given, it may be an educated guess. When several studies have been done, things narrow down a bit. Before long, a figure like $R_e=2.3$ is given with the confidence that it is most likely not 1.1, nor 15. An epidemiologist can say with confidence that $R_e=2.3$ give or take a few tenths of a percent. Of course, it is always possible that they missed something, like a lot of aymptomatic cases, which would inflate the figure for R_e.

Herein lies the problem. For the first month all of the above look the same to the general public and most politicians, with the possible exception of the last case with $R_e=5.8$. Having 10 cases or 100 cases or 1000 cases or even 10,000 cases in a couple months' time will look pretty much the same to the public and politicians, but not to the epidemiologist. The epidemiologist sees the future, whereas the public and politicians often cannot. The epidemiologist sees what a small change in a parameter looks like down the road in a couple months. Trust the epidemiologist - they know what they are talking about.

Stay well……..

Variables and Parameters

I(t)   number of infected people

S(t)   number of susceptible people

R(t)   number of recovered people

I_0 Initial number of infections

R_0 basic reproduction number

R_e effective reproduction number

r_e exponential growth rate

r_l logistic growth rate

τ duration of contagious period

K   maximum number of infections (logistic model)

c   proportionality constant, logistic differential equation

d   daily growth rate

x   fraction of susceptibles remaining after vaccinations

Formulas

$I(t) = I_0 e^{r_e t}$

$$I(t) = \frac{K}{1 + \frac{K - I_0}{I_0} e^{-r_l t}}$$

$r_l \approx r_e$

$r_l = cK; \; c = r_l/K$

$d = e^{r_e} - 1$

$r_e = \ln(d+1)$

$c = \ln(d+1)/K$

$c' = \ln(xd+1)/(xk)$

$$r_e = \frac{\ln(2)}{\text{time-to-double}}$$

New Infections Per Day=c(I)(K-I) (Logistic Model)

New Infections $= \dfrac{K}{\tau(S(0)+I(0)+R(0))} I(t)S(t)$ (SIR Model)

New Recoveries $= \dfrac{1}{\tau} I(t)$ (SIR Model)

Answers

Chapter 1: Simple Exponential and Logistic Models

1) $I(t) = 555e^{0.3307t}$ 2) $I(t) = \dfrac{3{,}000{,}000}{1 + 5{,}404.054 e^{-0.3309t}}$

Chapter 2: Making Predictions

1) a) $I = 264{,}863$ b) $t = 9.37$ days   2) a) $I = 243{,}865$ b) $t = 9.38$ days

3) $t = 19.3$ days

Chapter 3: What Mechanism Produces the Logistic S-Curve?

1) a) $c = 0.006666667$

b)

Number With Flu, I	Number Without Flu, K-I	Possible Interactions, I(K-I)	cI(K-I) infections/day	Days/infection	Cumulative Time (days)
1	14	14	0.09	10.71	
2	13	26	0.17	5.77	10.71
3	12	36	0.24	4.17	16.48
4	11	44	0.29	3.41	20.65
5	10	50	0.33	3.00	24.06
6	9	54	0.36	2.78	27.06
7	8	56	0.37	2.68	29.84
8	7	56	0.37	2.68	32.52
9	6	54	0.36	2.78	35.19
10	5	50	0.33	3.00	37.97
11	4	44	0.29	3.41	40.97
12	3	36	0.24	4.17	44.38
13	2	26	0.17	5.77	48.55
14	1	14	0.09	10.71	54.32
15	0	0	-	-	65.03

Total Time = 65.03 days

2) a) c20=0.2 -> c=0.01 (first 10 infections)

   c20=0.1 -> c=0.005 (second 10 infections)

b) Fill in the chart below.

Number With Flu, I	Number Without Flu, K-I	Possible Interactions, I(K-I)	cI(K-I) infections/ day	Days/ infection	Cumulative Time
1	19	19	0.19	5.26	0
2	18	36	0.36	2.78	5.26
3	17	51	0.51	1.96	8.04
4	16	64	0.64	1.56	10.00
5	15	75	0.75	1.33	11.56
6	14	84	0.84	1.19	12.90
7	13	91	0.91	1.10	14.09
8	12	96	0.96	1.04	15.19
9	11	99	0.99	1.01	16.23
10	10	100	1.00	1.00	17.24
11	9	99	0.50	2.02	18.24
12	8	96	0.48	2.08	20.26
13	7	91	0.46	2.20	22.34
14	6	84	0.42	2.38	24.54
15	5	75	0.38	2.67	26.92
16	4	64	0.32	3.13	29.59
17	3	51	0.26	3.92	32.71
18	2	36	0.18	5.56	36.63
19	1	19	0.10	10.53	42.19
20	0	0			52.72

c) Explain why this is an <u>asymmetric</u> S-curve. The first 10 days had a total time of 17.24 days, whereas the second 10 days had a total time of 52.72-17.24 =35.48 days.

Chapter 4: The Apex and S-Curve Basics

1)

Day	Total Infections	New Infections
0	1	
1	2	1
2	4	2
3	7	3
4	11	4
5	15	4
6	19	4
7	23	4
8	26	3
9	28	2
10	30	1

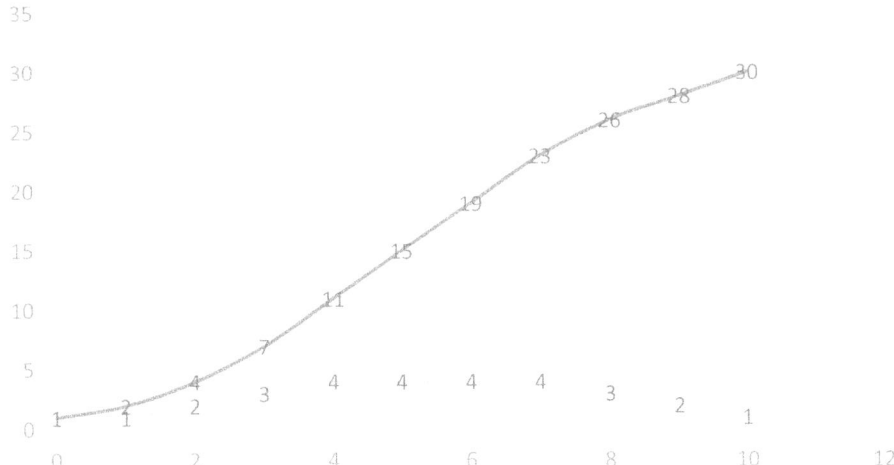

Infections and New Infections vs. Time

Chapter 5: Estimating the Logistic Growth Rate When $I_0 \ll K$

1)

Exponential Model	r_e	Estimated r_l	I_0	K	$I(t) = \dfrac{K}{1 + \dfrac{K - I_0}{I_0} e^{-r_l t}}$
$I(t)=12e^{0.1324t}$	0.1324	0.132	12	5,000	$I(t) = \dfrac{5000}{1 + \dfrac{5000 - 12}{12} e^{-0.132t}}$
$I(t)=125e^{0.0324t}$	0.0324	0.032	125	10,000	$I(t) = \dfrac{10{,}000}{1 + \dfrac{10{,}000 - 125}{125} e^{-0.032t}}$
$I(t)=150e^{0.1382t}$	0.1382	0.138	150	25,000	$I(t) = \dfrac{25{,}000}{1 + \dfrac{25{,}000 - 150}{150} e^{-0.138t}}$

2) Circle TRUE or FALSE. Use the rule given in the last sentence of this section.

a) $2 \ll 300$ **TRUE**   FALSE   b) $1 \ll 1000$ **TRUE**   FALSE

c) $3.2 \ll 59$ TRUE   **FALSE**   d) $22 \ll 23$ TRUE   **FALSE**

Chapter 6: Doubling Time and the Growth Rate

1) Fill in the following table:

Initial Infected Population	Days For Population To Double	$r_e = \dfrac{\ln(2)}{\text{time-to-double}}$	Model, $I(t) = I_0 e^{r_e t}$
100	5	0.1386	$I(t) = 100e^{0.1386t}$
200	4	0.1733	$I(t) = 200e^{0.1733t}$
1,000	10	0.0693	$I(t) = 1{,}000e^{0.0693t}$
1	7	0.0990	$I(t) = 1e^{0.0990tt}$

2) Fred has $10, then has $20 in 5 days, and then $30 in 10 days. His money does not double every 5 days, so growth is not exponential.

3) $I_0=25$ time-to-double = 5 days $r_e = \ln(2)/5 = 0.1386$ $I(t)=25e^{0.1386t}$

Chapter 7: What About R_0 and R_e?

1)

R_e	Fraction vaccinated for equilibrium $1-1/R_e$	Fraction without immunity, $x=1/R_e$	xR_e
1.1	0.0909	0.909	1
1.8	0.4444	0.5555	1
2.1	0.5238	0.4762	1
5.0	0.8	0.2	1
15.0	0.9333	0.0667	1

2)

Initial R_e	Percent With Herd Immunity	Percent Without Herd Immunity	New R_e
2.5	28%	100%-28%=72%	0.72x2.5=1.8
5.5	80%	100%-80%=20%	0.20x5.5=1.1
7.0	92%	100%-92%=8%	0.08x7.0=0.56

3) $1-1/3.2 = 0.6875 = 68.75\%$

4) $1-1/15 = 0.9333 = 93.33\%$

Chapter 8: r_e and the Daily Growth Rate

1)

Exponential Model	r_e	Daily Growth Rate, $d = e^{r_e} - 1$
$I(t)=40e^{0.15t}$	0.15=15%	$e^{0.15}-1=0.16=16\%$
$I(t)=80e^{0.20t}$	0.20=20%	$e^{0.20}-1=0.22=22\%$
$I(t)=21e^{0.25t}$	0.25=25%	$e^{0.25}-1=0.28=28\%$
$I(t)=120e^{0.18t}$	0.18=18%	$e^{0.18}-1=0.20=20\%$ (19.7%)
$I(t)=60e^{0.35t}$	0.35=35%	$e^{0.35}-1=0.42=42\%$

2)

Daily Growth Rate, d	$r_e = \ln(d+1)$	I_0	Model, $I(t) = I_0 e^{r_e t}$
20%	$\ln(0.20+1)=0.18$	20	$I(t) = 20e^{0.18t}$
35%	$\ln(0.35+1)=0.30$	15	$I(t) = 15e^{0.30t}$
0.12	$\ln(0.12+1)=0.11$	45	$I(t) = 45e^{0.11t}$
0.28	$\ln(0.28+1)=0.25$	150	$I(t) = 150e^{0.25t}$

3) a) $I(t) = \dfrac{2000}{1+99e^{-0.166t}}$ b) 23.57 c) 23.6

Chapter 9: Using R_e to find r_e

1)

R_e	τ days	d (infections/day)
15	3.0	5.0
2.8	5.5	0.509
1.2	4.0	0.3
1.8	5.0	0.36

2)

R_e	τ days	$r_e = \ln(R_e/\tau + 1)$
15	3.0	1.79
2.8	5.5	0.41
1.2	4.0	0.26
1.8	5.0	0.31

3) $r_e = \ln(1.8/7+1) = 0.23$, $I_0 = 12$, $I(t) = 12e^{0.23t}$

Chapter 10: The Effect of Social Distancing

1) We have 14 people with 1 infected. $R_e = 3.5$, $\tau = 5$ days, and $x = 0.5$ (50% removed due to social distancing).
a) Find c and c'. b) Estimate the total time before and after social distancing.

a) $c = \ln(R_e/\tau + 1)/K = \ln(3.5/5+1)/14 = 0.03790$

$c' = \ln(xR_e/\tau + 1)/(xK) = \ln(0.5(3.5)/5+1)/((0.5)(14)) = \ln(1.35)/7 = 0.04287$

b)

Number With Flu, I	Number Without Flu, K-I	Possible Interactions, I(K-I)	cI(K-I) infections/day	Days/infection
1	13	13	0.49	2.03
2	12	24	0.91	1.10
3	11	33	1.25	0.80
4	10	40	1.52	0.66
5	9	45	1.71	0.59
6	8	48	1.82	0.55
7	7	49	1.86	0.54
8	6	48	1.82	0.55
9	5	45	1.71	0.59
10	4	40	1.52	0.66
11	3	33	1.25	0.80
12	2	24	0.91	1.10
13	1	13	0.49	2.03
14	0	13	0.49	2.03

Total Time = 11.99 days

c)

Number With Flu, I	Number Without Flu, K-I	Possible Interactions, I(K-I)	c'I(K-I) infections/day	Days/infection
1	6	6	0.25722	3.887723
2	5	10	0.4287	2.332634
3	4	12	0.51444	1.943861
4	3	12	0.51444	1.943861
5	2	10	0.4287	2.332634
6	1	6	0.25722	3.887723
7	0	0		

Total Time = 16.33 days

Chapter 11: The SIR Model Simply Explained

1)

$$\frac{R_e}{\tau(S(0)+I(0)+R(0))} = \frac{2.8}{5(490+10+0)} = 0.00112 \qquad \frac{1}{\tau} = \frac{1}{5} = 0.2$$

day	S(t)	I(t)	R(t)	New Infections $\frac{R_e}{\tau(S(0)+I(0)+R(0))} \cdot S(t)I(t)$	New Recoveries $\frac{1}{\tau}I(t)$
0	490	10	0	0.00112x490x10=5	0.2x10=2
1	485	13	2	7	3
2	478	17	5	9	3
3	469	23	8	12	5
4	457	30	13	15	6
5	442	39	19	19	8

www.ingramcontent.com/pod-product-compliance
Lightning Source LLC
Chambersburg PA
CBHW080937220526
45465CB00008BA/3074